SPOTLIGHT ON NATURE

WOLF

LAURA PURDIE SALAS

CREATIVE EDUCATION · CREATIVE PAPERBACKS

Published by Creative Education and Creative Paperbacks
P.O. Box 227, Mankato, Minnesota 56002
Creative Education and Creative Paperbacks are imprints
of The Creative Company
www.thecreativecompany.us

Design by Rhea Magaro; production by Beeline Media & Design, Inc.
Art direction by Tom Morgan
Printed in the United States of America

Photographs by Alamy (Adam Jones / DanitaDelimont, H. Mark Weidman Photography, imageBROKER/W. Rolfes, Glenn Nagel), Getty Images (Jim Cumming, Adria Photography, PHOTO 24, Martin Ruegner, Daniel Parent, Raimund Linke, Stan Tekiela Author / Naturalist / Wildlife Photographer, Picture by Tambako the Jaguar, Patrick J. Endres, imageBROKER/Stefan Huwiler, John Morrison), Public Domain (NPS/Doug Smith, USFWS), Shutterstock (photomaster, Ramon Carretero, AN NGUYEN, evanesa, Giedriius)

Library of Congress Cataloging-in-Publication Data
Names: Salas, Laura Purdie, author.
Title: Wolf / Laura Purdie Salas.
Description: Mankato, Minnesota : Creative Education and Creative Paperbacks, [2025] | Series: Spotlight on nature | Includes bibliographical references and index. | Audience: Ages 10-13 | Audience: Grades 4-6 | Summary: "Howl along with the wolf's world with our new book for mid-level readers. An engaging narrative of a wolf family, captivating images, infographics, milestones, and resources make it perfect for young nature enthusiasts"-- Provided by publisher.
Identifiers: LCCN 2023059510 (print) | LCCN 2023059511 (ebook) | ISBN 9798889891970 (library binding) | ISBN 9781682775707 (paperback) | ISBN 9798889892052 (ebook)
Subjects: LCSH: Gray wolf--Juvenile literature. | Gray wolf--Life cycles--Juvenile literature. | Wolves--Juvenile literature. | Wolves--Life cycles--Juvenile literature. | Gray wolf--Minnesota--Juvenile literature.
Classification: LCC QL737.C22 S26 2025 (print) | LCC QL737.C22 (ebook) |
DDC 599.773--dc23/eng/20240129
LC record available at https://lccn.loc.gov/2023059510
LC ebook record available at https://lccn.loc.gov/2023059511

CONTENTS

MEET THE FAMILY

GRAY WOLVES

of the Minnesota Northwoods

The Minnesota Northwoods is a vast region of crystal-clear lakes and endless forests in the northern United States. The ground is carpeted by the leaves and needles of spruce, pine, birch, and maple trees. Boulders, dragged by ancient glaciers, litter the land. There is abundant wildlife, including loons, eagles, foxes, raccoons, black bears, and moose.

On a cold, sunny, May day, seven gray wolves gather. The pack is led by an **alpha** male and female. There are also four of their young offspring from previous years. The last wolf is an adult. She's the sister of the alpha female. The family is about to get larger.

In March, as winter loosened its grip, the alpha pair mated. It's now nine weeks later, and the alpha female is restless. She stays close to the den, which has been hollowed out under a tree. She knows there is room inside for her and the coming pups.

CLOSE-UP

Eye color

Pups open their eyes when they are about two weeks old. All gray wolf pups have blue eyes. Gradually, their eye color changes. Most adult wolves have gold, hazel, light brown, or pale green eyes.

CHAPTER ONE

LIFE BEGINS

Wolves are large, wild members of the dog family. The **DNA** of wolves and pet dogs is 99.9 percent the same. There are many **species** and **subspecies** of wolves. The largest can grow to 175 pounds (79 kilograms). The gray wolf is the best-known species. Two others are the red wolf and the Ethiopian wolf. Another animal used to be called the African golden jackal. Scientists studied its DNA and found that it was a wolf, so they renamed it the African golden wolf.

Wolves have adapted to live in many different habitats. They can live in forests, mountains, grasslands, tundra, and deserts. Gray wolves live in parts of North America, Europe, and Asia. Red wolves live only in a tiny region of forests and wetlands in the U.S. state of North Carolina. They're almost extinct. African golden wolves live in northern African grasslands. Ethiopian wolves live only in the country of Ethiopia in grasslands high in the mountains.

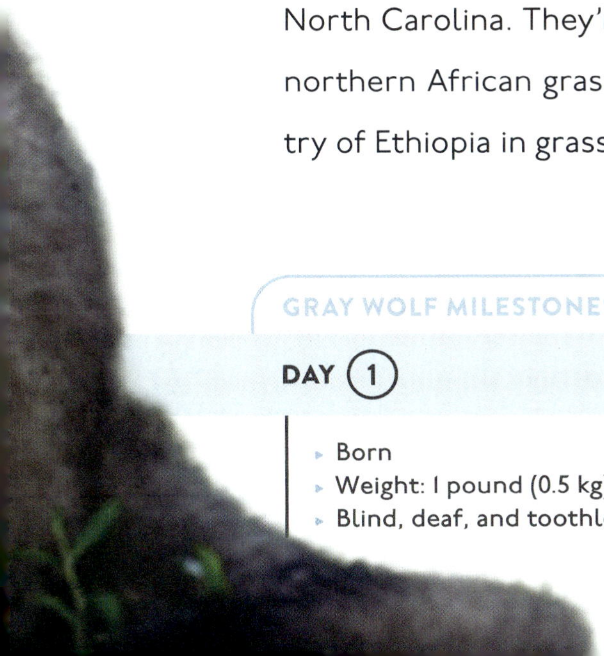

GRAY WOLF MILESTONES

DAY 1

- Born
- Weight: 1 pound (0.5 kg)
- Blind, deaf, and toothless
- Can't control body temperature
- Drinks milk 4 to 5 times each day

Wolves are carnivores. They are apex predators. Except for humans, no other animal hunts adult wolves. Wolves eat large mammals such as deer and moose. They will also eat small mammals, fish, and reptiles. They have even been known to eat fruit. When prey is scarce, wolves may change their eating habits.

Wolves live in packs, which are like families. Most packs have six to 10 wolves. If there is plenty of prey, larger packs can form. Each pack lives in a home territory. The area can be as small as 75 square miles (194 square kilometers) and as large as 3,000 square miles (7,770 sq. km). Territories are smaller where there are plenty of prey animals.

CLOSE-UP

Teeth

Adult wolves have 42 teeth that are adapted for eating meat. Long canine teeth grip prey. Sharp carnassial teeth can slice meat off bones. Back molars chew up tough meat.

FEATURED FAMILY

Welcome to the World

Inside the den, the female gray wolf shifts and stretches. She howls in discomfort as the first pup is born. The mother licks it clean, which helps the pup's circulation. Exhausted, the mother rests until the next pup is born. More pups follow. The mother gently lifts the small, blind pups with her jaws. Their coats are thin. She moves them close to her body for warmth. Outside, the others soon hear five pups chirping and squealing. The pack keeps its distance from the new arrivals. For about a month, the mother and pups will stay alone in the den where they are safe and warm.

Pack members have strong social bonds. Wolf packs hunt together, play together, and take care of each other. They communicate with sounds, scents, and body positions.

Often, only the alpha male and female mate. This helps ensure there's enough food for everyone. Once a year, the alpha female gives birth to a litter of four to seven pups. A gray wolf pup weighs just one pound (0.5 kg). It is helpless and totally dependent on its mother. For several weeks after birth, the mother and pups stay in the den.

(2) WEEKS

- Weight: 3.5 pounds (1.6 kg)
- Eyes open, but can't see well
- Tries to stand

(3) WEEKS

- First baby teeth appear
- Sight has improved
- Hearing has developed

CLOSE-UP

Howling at the moon?

Contrary to legend, wolves don't howl at the full moon. Wolves are active at night, so that's when people hear them. However, researchers haven't found a link between howling and the phase of the moon. Wolves do raise their snouts to howl. That's probably just to help the sound carry a long distance.

FEATURED FAMILY

First Meal

In the chilly darkness, the mother gray wolf and her pups snuggle together. The newborns are toothless and cannot eat solid food. Drinking their mother's milk is the only skill they have. For a few weeks, while the pups are nursing, the alpha male brings meat for his mate. Rich milk helps the pups grow strong. They are soon tumbling around the den. The oldest is a female. One day, she pokes her snout outside the den. She's ready to explore.

DOG FAMILY SIZES

BREED	HEIGHT AT SHOULDER
GRAY WOLF	29 inches (74 centimeters)
MEXICAN WOLF	28 inches (71 cm)
COYOTE	18 inches (46 cm)
PUG	12 inches (30 cm)
CHIHUAHUA	7 inches (18 cm)

4 WEEKS

- Weight: 5 to 6 pounds (2.3 to 2.7 kg)
- Leaves den and plays nearby
- Eats regurgitated meat from adults
- Makes high-pitched howls

CLOSE-UP

Rallies

When wolves howl together, it's called a rally. A rally shows that a pack is a family and that they are stronger together. Wolf howls can be heard three to seven miles (4.8 to 11.3 km) away.

CHAPTER TWO

EARLY ADVENTURES

Wolf pups leave the protection of the den when they are a few weeks old. The entire pack helps care for them. At least one babysitter wolf stays with the pups while the other wolves hunt. When the hunters return, they **regurgitate** food for the hungry pups.

Young pups are rambunctious. They climb all over the other wolves, including the adults. They bite or gnaw on legs and tails. Luckily, adult wolves are patient and caring with the pups.

After the wolf pups are **weaned**, the pack leaves the den behind. The alpha female leads her pups to the pack's first **rendezvous site**. This is a kind of home base for the pack that is near a water source. The pups stay at the site with a guard wolf when the pack leaves to hunt. Over the summer months, the pack will use a series of rendezvous sites. The growing pups wander farther and farther away from

6 WEEKS

- Has adult hair around face
- Play-fights with siblings
- Wanders away from den, but finds the way back

2 MONTHS

- Weight: 16 pounds (7.3 kg)
- Weaning off mother's milk
- Has large feet and head
- Stays behind at rendezvous site when pack hunts

each site. Soon, they may ramble away for two to three miles (3 to 5 km) before finding their way back. The pack will keep using rendezvous sites until the pups are old enough to join the hunt.

Wolf pups face two major dangers. The first is starvation. In some years, prey is scarce. The pack might not be able to hunt enough to feed everybody. The second danger is hunting. In some places, it is legal to hunt wolves. Only about half of wolf pups survive to adulthood.

When there's enough food, the pack thrives. The wolves fill their days and nights with hunting, eating, and playing. Their haunting howls travel through the woods. They sing together as one big family.

CLOSE-UP

Toes

Wolves walk and run on their toes. Soft pads on the undersides of their toes act as shock absorbers when they run. This protects their bones and joints. It also makes a wolf's run light and springy.

FEATURED FAMILY

Look Who's Leaving the Den

The gray wolf pup emerges from the den into the bright sunlight. The other four pups follow their sister. They're bigger and stronger now. They can see and hear, but they still have no teeth. The pack sniffs and licks the pups. The pups chirp, squeal, and whine. As they discover the world, they run, dig, and wrestle. When most of the pack leaves to hunt, one or two wolves stay behind to babysit. When the hunters return, they regurgitate part of their meal on the ground. The female pup slurps up the soupy mess.

(3) MONTHS

- Weight: 25 pounds (11.4 kg)
- Travels on hunts with rest of pack
- Observes hunting behaviors
- Eats meat

CLOSE-UP
Nose

An excellent sense of smell is crucial for hunting. As a wolf walks, it keeps its nose to the ground. It can smell a moose 300 yards (274 meters) away. A wolf has about 280 million scent receptors. Humans have only about 500.

FEATURED FAMILY

Give It a Try

Now 12 weeks old, the female gray wolf pup no longer drinks milk. Her teeth have grown in. When the pack hunts, she and her siblings follow. She watches the alpha male chase a moose. He grips it in his strong jaws and hangs on. Other wolves join in, and the pack works together to drag down the moose. The pack begins to feed, but the pup hangs back. She isn't sure what to do. The alpha pair nudges the others aside to make room for her. She and her siblings crowd in. She rips off strips of meat and eats until her belly is very full.

WOLF PACKS HUNT, EAT, and PLAY TOGETHER.

4 MONTHS

- Loses baby teeth
- Develops size, strength, and speed needed for hunting

CLOSE-UP

Walking

Many people think wolves spend all their time hunting. Actually, wolves spend most of their time walking. They walk up to eight hours every day. They often cover 30 miles (48 km) in a single day.

CHAPTER THREE

LIFE LESSONS

Gray wolves hunt in packs. A single gray wolf could hunt alone and survive on small animals. So why do wolves hunt in groups? Experts believe it is a strategy that benefits pups. Young wolves can't hunt for themselves until they are about a year old. Until then, adults have to catch extra food to feed the pack's pups.

Hunting takes time and energy. Wolves might eat only once a week. When there's a kill, they eat a large meal. Hunting is also dangerous. Wolves often hunt animals 10 times their own size. To catch a moose, a wolf will grip its rump and stay attached. The pack joins the battle. The enormous moose is powerful. It kicks. It swerves and swings to fling the wolves off. A wolf might crash into a tree or land hard on the ground. Sometimes, moose kill wolves.

6 MONTHS

- Weight: 64 pounds (29 kg)
- Reaches about 80 percent of adult size
- Growth slows
- Actively participates in hunting

1 YEAR

- Now an excellent hunter
- Now past the age when pups are most vulnerable to death and disease
- Helps care for new brothers and sisters

Packs are close families, but individual wolves come and go. Wolves usually mate for life. However, their pups don't stay with the same pack forever. When they are one or two years old, young adult wolves go off on their own. Two-year-old wolves are sexually mature. They search for a mate. A mated pair and their pups start a new pack.

Wolves influence the environment. A healthy wolf population keeps the population of prey animals in check. This impacts the land. Large deer populations overgraze the land so that new trees never get very tall. When there are more wolves, there are fewer deer. When there are fewer deer, forests can grow.

FEATURED FAMILY

This Is How It's Done

The female gray wolf pup is learning to communicate. She yips when she wants to play. Her brother chases her through the woods. When he won't stop pouncing on her, she growls to say, "Knock it off." When he slinks away and tucks his tail between his legs, she feels strong. She holds her head and tail high. A smell catches her attention, and she wanders away to investigate. She finds a scent mark left by the alpha male. It means she's at the edge of the pack's territory. She howls to let her family know where she is.

CLOSE-UP

Hunting strategies

Wolves chase their food. Once a wolf grips a prey animal, other members of the pack pile on and try to drag the animal down. Wolves also have more advanced techniques. While one wolf chases the prey, others will block the animal's escape routes.

2 YEARS

- Weight as fully-grown female: 80 pounds (36 kg)
- Leaves birth pack
- Finds mate

3 YEARS

- Gives birth to first litter of pups
- Becomes alpha female of new pack

Wolves also impact forests by preying on beavers. Beavers would rather gnaw down leafy trees than evergreens. If beavers have no predators around, they travel far from their lodges to find their favorite trees. However, when wolves live nearby, beavers play it safe. They stay closer to their lodges and gnaw down whatever trees are closest. This protects leafy trees in the region.

CLOSE-UP

Food

Wolves don't waste food. When they make a kill, they eat the animal's muscles, tendons, skin, and many bones. They leave behind only very large bones, some hair, and some stomach contents.

Most wolves live for six to eight years in the wild. Many die sooner. Some can be as old as 13.

Wolves are feared because they are predators, but they do not deserve their bad reputation. They are playful and loyal. They are wild, shy, and fierce. They are an important part of healthy forests and food chains. Wolves are beautiful symbols of nature.

FEATURED FAMILY

Practice Makes Perfect

The first-born gray wolf is now a young adult. She is grown, but wolves never outgrow the desire to play. Playing together strengthens the bonds between members of the pack. The female lowers her front paws, raises her rump, and wags her tail. An older wolf plays chase with her. Later, she wrestles with her siblings. They use bones and sticks as toys. The female struts around with a fox bone. She tosses it into the air and then grabs it back. When a sibling dashes in and steals it, she chases him to get her toy back.

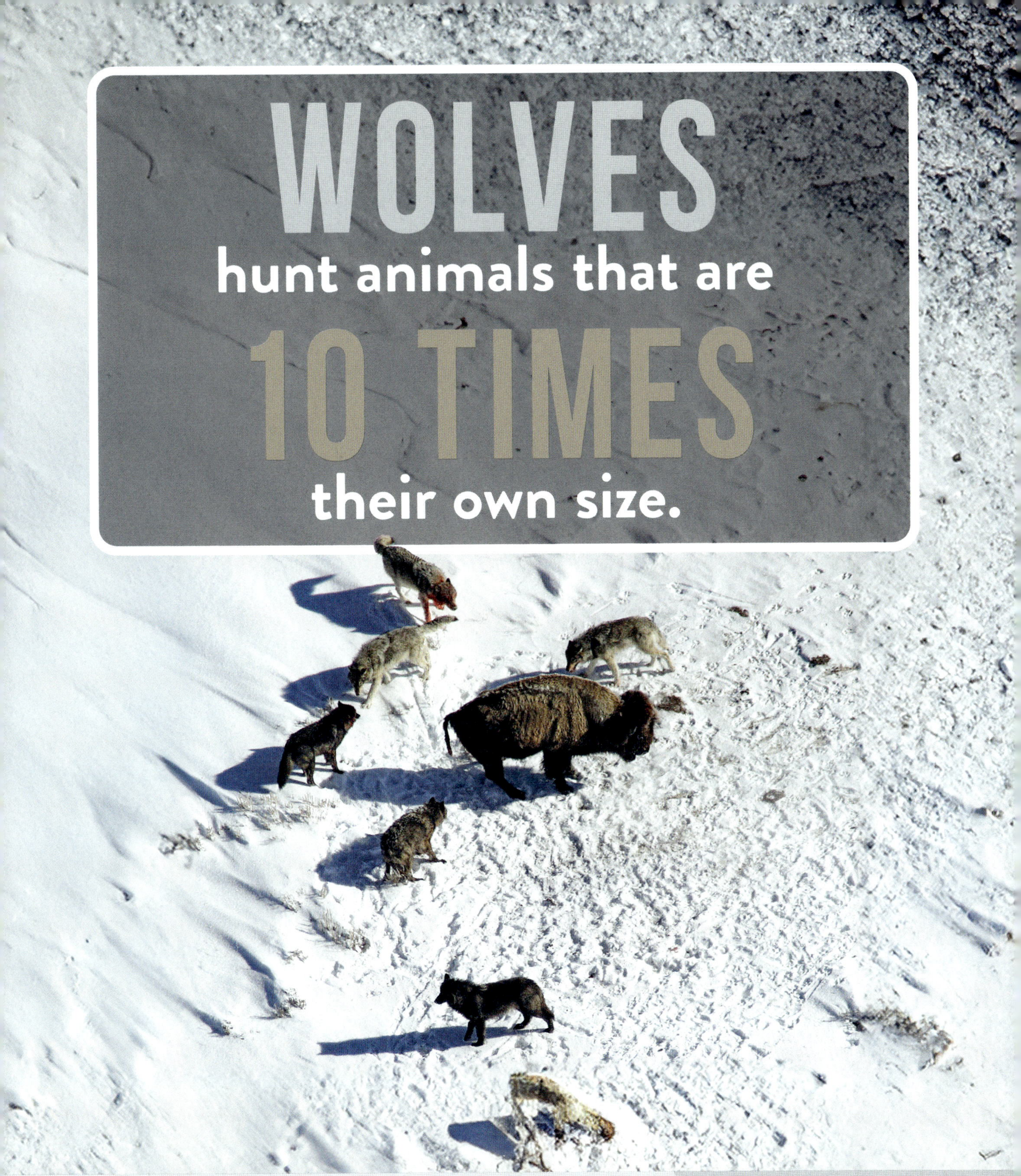

WOLVES hunt animals that are 10 TIMES their own size.

(8) YEARS

- Gives birth to last litter of pups
- Teeth are worn down, making hunting and feeding more difficult
- End of life

CHAPTER FOUR

HELPING WOLVES SURVIVE

The biggest threat to wolves is human hatred. Many Native American cultures celebrate the strength and beauty of wolves. That's a rare attitude. Most people associate wolves with villains from fairy tales such as "Little Red Riding Hood." Wolves rarely get to be heroes.

Hundreds of years ago, two million gray wolves lived in North America. By the 1950s, hunters had killed almost all of them. The Endangered Species Act of 1973 protected wolves. Their numbers grew. In 1995, wolves were brought back to Yellowstone National Park, where they hadn't lived for almost 100 years. By 2020, about 7,500 wolves lived in the continental United States.

Then, the laws changed again. Wolves lost some of their protection. Some U.S. states allow wolf hunting. In 2021, the state of Idaho passed laws allowing 90 percent of its wolves to be killed. Hunters can use night-vision goggles and shoot from helicopters. These methods are illegal for hunting most other animals.

Some people who live near wolves say they are scared of being attacked. Some hunters say that wolves kill all the elk and other big game. Some farmers say that wolves might kill their cows or sheep. Research studies do not support these claims. They show that wolves make elk populations healthier by killing older and weaker animals. They show that wolves rarely attack humans or livestock.

As humans build more roads, farms, and houses, wolves lose their habitats. The North Carolina red wolf is critically endangered. Fewer than 50 animals remain. However, some wolves are finding ways to survive. Overall, gray wolves have a stable population. In the early 2020s, there were about 18,000 gray wolves living in the United States.

People in **conservation** groups work to protect wolves. They set aside land where wolf packs can live. They put tracking devices on red wolves to study them and to discourage hunters. They take red wolf pups born in captivity and place them with wild red wolves. Many zoos and wildlife centers care for wolves. They allow visitors to get close to wolves and learn about their importance to the environment.

Humans around the world admire, love, or despise wolves. Only through education and science can more people learn to appreciate these amazing animals.

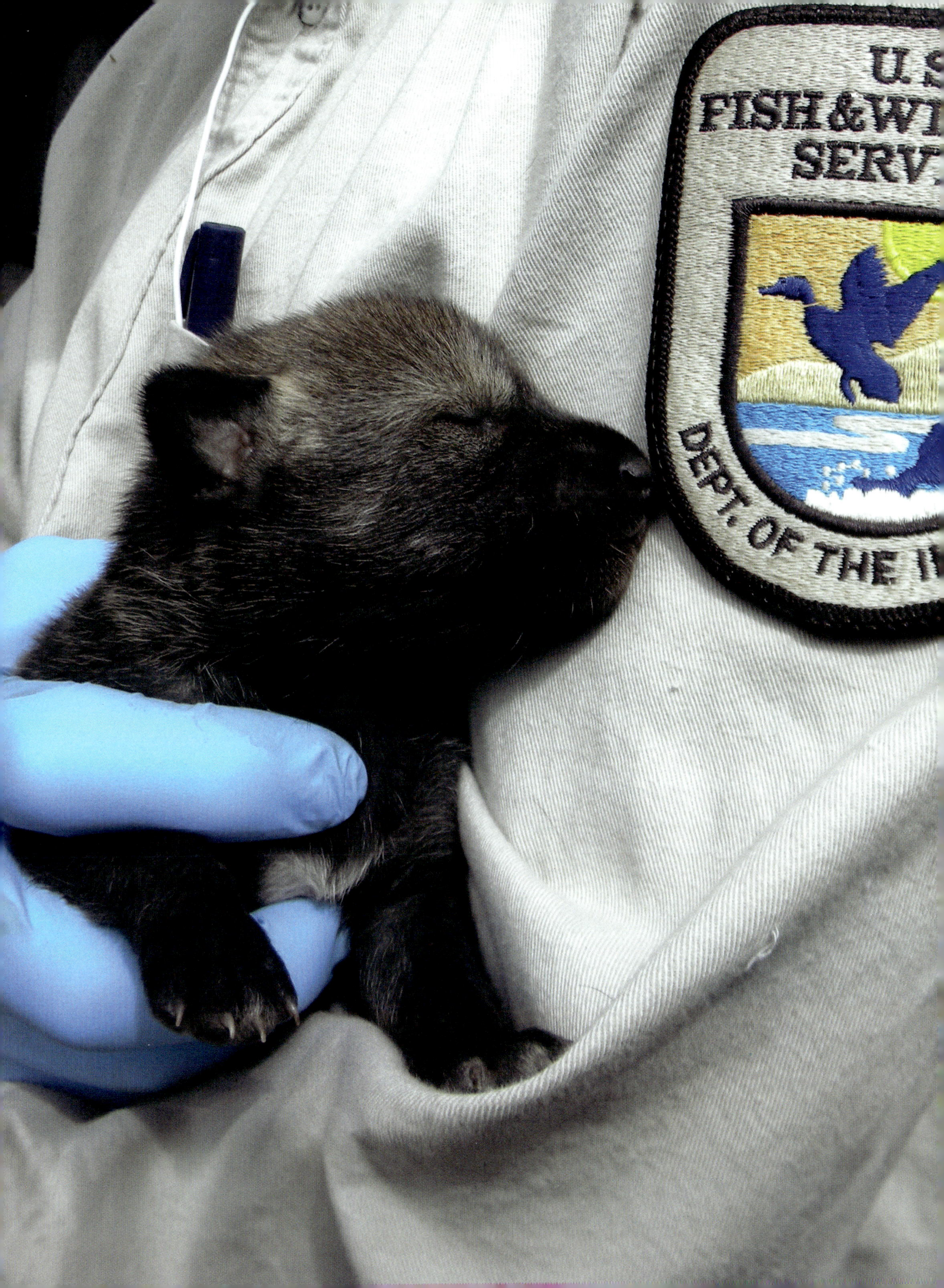
DEPT. OF THE

FAMILY ALBUM

SNAPSHOTS

Arctic wolves have white fur all year.

Red wolves are the only wolf species that lives only in the United States.

Rabies kills many **Ethiopian wolves**. To help protect them, scientists put out goat meat injected with rabies vaccines.

Mackenzie Valley wolves might be the largest wolves. The heaviest one on record was 175 pounds (79 kg).

Gray wolves don't hibernate. To stay warm, a mother and her pups stay in a den for a few weeks right after the pups are born.

Eastern wolves were renamed **Algonquin wolves** in 2016. As scientists learn more, they often rename or reclassify wolf species and subspecies.

Mexican gray wolves are often called *el lobo*, which means "the wolf" in Spanish.

The **coastal wolves** of Vancouver Island eat seafood. They catch salmon, otters, and seals.

Small **Arabian wolves** weigh about 40 pounds (18 kg). Having large ears and a thin coat helps these animals survive in their desert habitat.

Himalayan wolves live in cold, mountainous regions. They have extra-thick coats and are sometimes called woolly wolves.

Dingoes are wild dogs in Australia. Most scientists believe that dingoes are a subspecies of wolves.

Greenland wolves usually live in packs of just three animals.

WORDS to Know

alpha	one who leads; a wolf pack has an alpha female and an alpha male
apex predators	animals at the top of the food chain that are not hunted by any other wild animals
carnivores	animals that eat mostly meat
circulation	the act of blood moving through the body to spread warmth and energy
conservation	planned management of a resource such as land or wildlife to preserve and protect it
DNA	abbreviation for *deoxyribonucleic acid*; molecules inside a living thing that hold its genetic information and determine its species and characteristics
regurgitate	vomit
rendezvous site	a meeting place that is agreed on by all members of a group
species	a group of living things that have shared characteristics and that are able to reproduce with one another
subspecies	types or varieties of living things within a species
weaned	able to eat solid food; not dependent on a mother's milk

LEARN MORE

Books

Carson, Mary Kay. *The River That Wolves Moved: A True Tale from Yellowstone*. Ann Arbor, Mich.: Sleeping Bear Press, 2022.

Furstinger, Nancy. *We Need Wolves*. Mendota Heights, Minn.: North Star Editions, 2019.

Patent, Dorothy Hinshaw. *When the Wolves Returned*. New York: Bloomsbury Children's Books, 2008.

Websites

"Just for Kids." International Wolf Center.
https://wolf.org/wolf-info/wild-kids/

"The Language of Wolves." Living with Wolves.
https://www.livingwithwolves.org/about-wolves/language/

"Webcams." Wolf Conservation Center.
https://nywolf.org/meet-our-wolves/webcams/

Documentaries

Douglas, David. *IMAX Wolves*. IMAX and Go Planet, 2010.

Dutcher, Jamie and Jim. *Living with Wolves*. Discovery Channel, 2007.

Goetzl, Oliver. *Nature: Arctic Wolf Pack*. PBS/Thirteen Productions, 2018.

Note: Every effort has been made to ensure that any websites listed above were active at the time of publication. However, because of the nature of the Internet, it is impossible to guarantee that these sites will remain active indefinitely or that their contents will not be altered.

Visit

CALIFORNIA WOLF CENTER

Take a tour to meet the ambassador wolf packs and learn about conservation efforts.

Tall Pine Rd. & K Q Ranch Rd.

Julian, CA 92036

INTERNATIONAL WOLF CENTER

Visit to learn about wolves, their relationship to the land, and their relationship with humans.

1396 Highway 169

Ely, MN 55731

WOLF CONSERVATION CENTER

Attend a program and meet ambassador wolves.

7 Buck Run St.

South Salem, NY 10590

INDEX